THE ELECTROMAGNETIC SPECTRUM

UNDERSTANDING AND USING UV AND INFRARED RADIATION

JONATHAN AND MARIEL BARD

Enslow PUBLISHING

Published in 2022 by Enslow Publishing, LLC.
101 W. 23rd Street, Suite 240, New York, NY 10011

Library of Congress Cataloging-in-Publication Data

Names: Bard, Jonathan, author. | Bard, Mariel, author.
Title: Understanding and using UV and infrared radiation / Jonathan and Mariel Bard.
Description: New York, NY : Enslow Publishing, 2022 | Series: The electromagnetic spectrum | Includes bibliographical references and index. | Audience: 7–12.
Identifiers: LCCN 2019014894| ISBN 9781978514959 (library bound) | ISBN 9781978514942 (pbk.)
Subjects: LCSH: Ultraviolet radiation—Juvenile literature. | Infrared radiation--Juvenile literature.
Classification: LCC QC459 .B37 2022 | DDC 535.01/2—dc23
LC record available at https://lccn.loc.gov/2019014894

Printed in the United States of America

To Our Readers: We have done our best to make sure all websites in this book were active and appropriate when we went to press. However, the author and the publisher have no control over and assume no liability for the material available on those websites or on any websites they may link to. Any comments or suggestions can be sent by email to customerservice@enslow.com.

Photos Credits: Cover dem10/E+/Getty Images; p. 5 VectorMine/Shutterstock.com; p. 8 udaix/Shutterstock.com; p. 10 MicroOne/Shutterstock.com; p. 12 Shawshots /Alamy Stock Photo; p. 14 magnetix/Shutterstock.com; p. 16 Photos.com /Getty Images; p. 20 elenabsl/Shutterstock.com; p. 22 Cultura RM Exclusive /Joseph Giacomin/Getty Images; p. 25 Jingjits Photography/Shutterstock.com; p. 27 NASA/JPL-Caltech; p. 29 Francisco Javier Ramos Rosellon/Alamy Stock Photo; p. 31 Science History Images/Alamy Stock Photo; p. 35 diy13/Shutterstock.com; p. 40 Marccophoto/iStock/Getty Image; p. 44 Science Photo Library/Alamy Stock Photo; p. 46 Karen Bleier/AFP/Getty Image; p. 47 John B. Carnett/ Popular Science/Getty Image; cover background, interior pages (faceted color pattern) s_maria/Shutterstock.com; interior pages (wave graphic) MagicMary/ Shutterstock.com.

CONTENTS

INTRODUCTION

The electromagnetic spectrum includes the entire range of electromagnetic radiation. This radiation includes all forms of light—not only ultraviolet and infrared waves, but also visible light, radio waves, microwaves, X-rays, and gamma rays. Light is also called electromagnetic radiation because it is a wave of electric and magnetic fields.

Electromagnetic energy travels in waves. Picture ocean waves: They come in different sizes, travel at different speeds, and crash into the shore with different amounts of force. Waves of radiation are very similar. They travel at different frequencies; they have different wavelenghts, or distances between their crests; and they carry different amounts of energy. A diagram of the electromagnetic spectrum shows the different wavelengths and frequencies of light energy. These different energies are all around us, but most electromagnetic radiation goes unseen by the human eye. What people *can* see is called visible light.

Humans are capable of seeing all sorts of colors, from purple to red and many colors in between. The ability to detect these colors comes from specialized cells in the eyes, called photoreceptor cells. There are two types of photoreceptors—rods and cones—in the retina of the eye. Rod cells detect black and white, and they help people with depth perception (which is how you know when and where to reach out your hand when someone throws you a ball). Cone cells, on the other hand, detect colors.

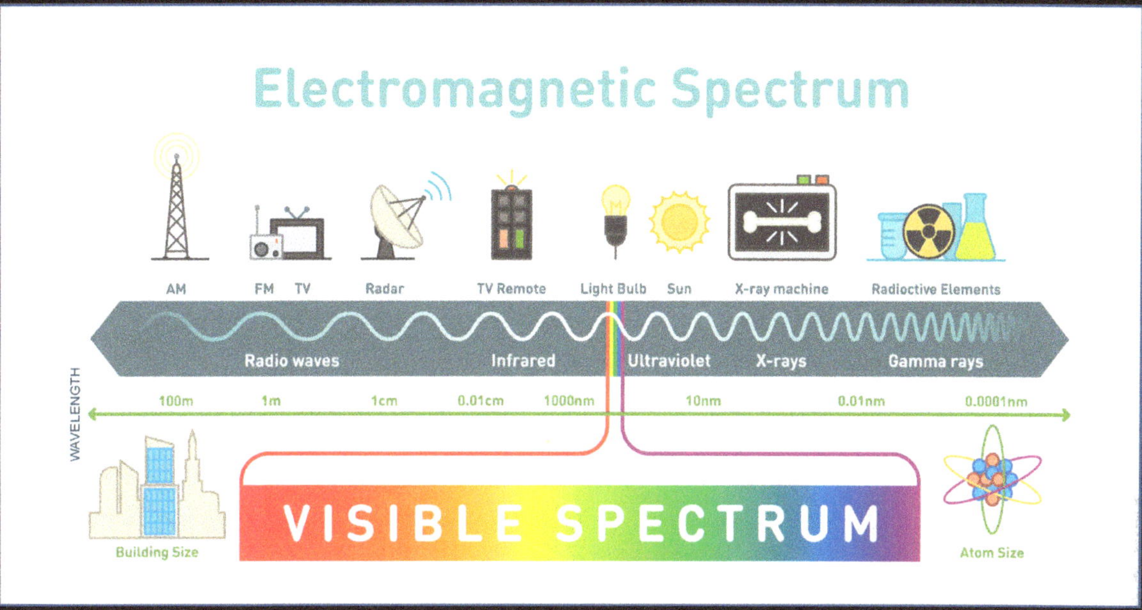

The electromagnetic spectrum arranges all the wavelengths of radiation in order from longest to shortest. The shorter the wavelength, the more dangerous it can be for living things.

Humans have three different versions of cone cells. Each version detects different parts of the visible-light spectrum. Another way to think of the visible-light spectrum is as a rainbow—red, orange, yellow, green, blue, indigo, and violet. Each color has a different wavelength.

The visible light that the human eye can detect is only a small section of a much bigger range of light energy that exists on the electromagnetic spectrum. And the human body alone cannot detect everything on this spectrum—we can't even detect *most* of it! Visible light is just one of the forms of electromagnetic radiation. Other forms include X-rays, which people use every day in medical offices and hospitals, and radio waves, which people use to broadcast news, sports, and the latest pop songs around the world. Ultraviolet (UV) radiation and infrared (IR) radiation are visible light's two closest neighbors on the electromagnetic spectrum. These have the most similar properties to visible light. Although humans cannot see them, other animals can. Reindeer, for example, are able to see traces of UV radiation in urine left by predators. They also use UV to find lichen to eat. And believe it or not, scorpions glow under UV light.[1] Some animals can see in IR, too. Mosquitoes use IR to find heat sources off of which they can feed, while snakes use IR to see in the dark.

Humans, as you may have noticed, lack these special abilities. But thanks to Sir William Herschel and his discovery of infrared radiation in 1800, people have been able to use infrared radiation in all sorts of technologies, from remote controls to complex telescopes. In 1801, shortly after Herschel's discovery, Johann Wilhelm Ritter discovered ultraviolet radiation. Scientists have continued to work to not only understand the damaging effects of UV radiation but also find ways people might benefit from UV light, such as using it to clean medical equipment in hospitals.

THE BASICS OF LIGHT ENERGY

CHAPTER ONE

Just like waves in the ocean, light waves can be measured in three ways: amplitude, frequency, and wavelength. The amplitude is the height of the wave, measured as half the distance from its highest point (called its crest or peak) to its lowest point (called its trough). A really big tidal wave in the ocean has a much larger amplitude than a wave made by a rowboat paddling on a pond. The frequency is the number of wave crests that pass a certain point in a given amount of time. On windy days, more wave crests might come to shore from the ocean, meaning they are more frequent or have a higher frequency. Frequency is measured in hertz, which is the number of wave crests per second. The unit hertz is named after German physicist Heinrich Hertz, who was the first to demonstrate that both light and heat are forms of electromagnetic radiation. The wavelength is the distance between each of the wave's peaks, and this is measured in meters. The wavelength and the frequency of light energy are inversely related, which means as the wavelength increases (gets longer), the frequency decreases.

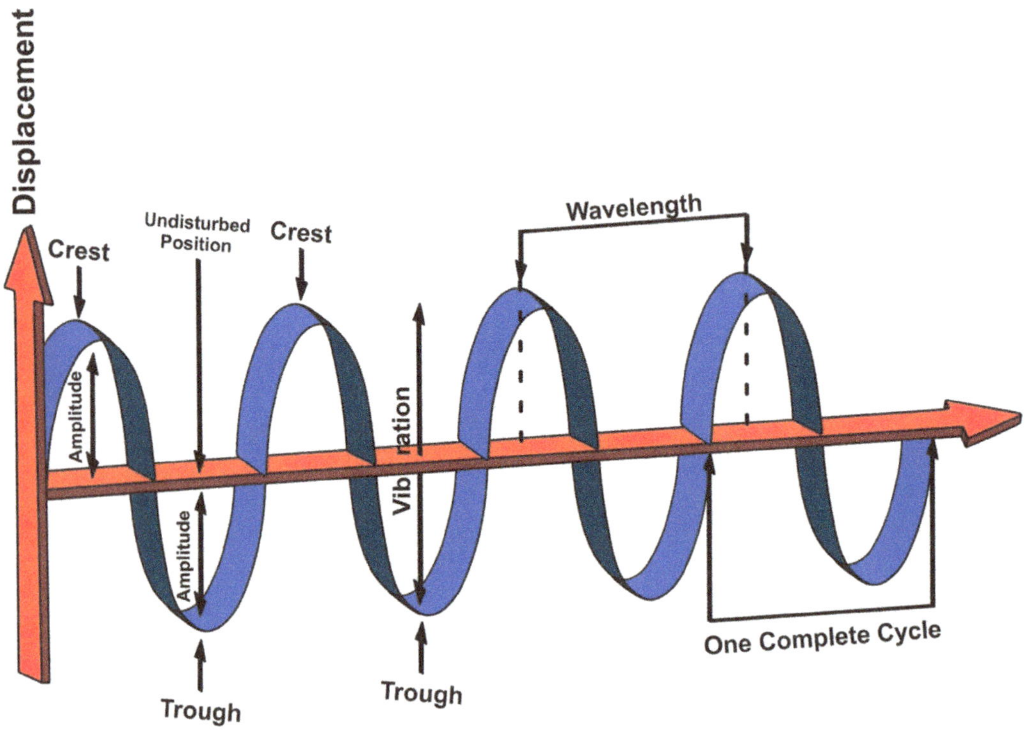

The crest of a wave is the highest point, and the trough is the lowest point. The distance between either of these points and the midpoint of the wave is the amplitude.

Wavelengths and the Electromagnetic Spectrum

Light energy comes in many different wavelengths and frequencies. This gives the full range of wavelength size incredible flexibility. The largest wavelengths are found on the radio-wave portion of the spectrum. These waves can be hundreds of meters in size. In contrast, the smallest wavelengths belong to gamma rays, which are at the opposite end of the spectrum and are less than

0.01 nanometers (nm).[1] That's very small—there are one billion nanometers in a meter! This massive range of electromagnetic frequencies and wavelengths gives each section of the electromagnetic spectrum interesting and unique properties.

The visible light that eyes can see has a wavelength between approximately 400 nm and 700 nm. Each color is found within this region of the electromagnetic spectrum. For example, the color red is found at one end of the visible-light spectrum, with wavelengths between 625 and 700 nm, while the color violet is found at the other end of the spectrum, with wavelengths between 380 and 435 nm.[2] For humans, each of the three types of cone-cell photoreceptors in the eyes specializes in a different range of visible-light wavelengths, and this is what allows humans to see all the colors of the rainbow. If you or someone you know is color blind, it's usually because one of these types of cone-cell photoreceptors cannot detect the wavelength it's supposed to detect. For those who are red-green color-blind, for example, their eyes have trouble distinguishing the colors found within wavelengths between 500 nm and 700 nm.[3]

Wavelengths that are a bit longer than red visible light are part of the infrared radiation band of the electromagnetic spectrum. Infrared occupies a wavelength range of 750 nm to 1 millimeter. These wavelengths are longer and are found at lower frequencies (between approximately 3×10^{12} and 4.3×10^{14} hertz).[4] The name "infrared" comes from the Latin word *infra*, meaning "below." Therefore, "infrared" means lower than the red portion of the visible-light spectrum.

Waves with shorter wavelengths and higher frequencies than violet visible light fall into the ultraviolet radiation band.

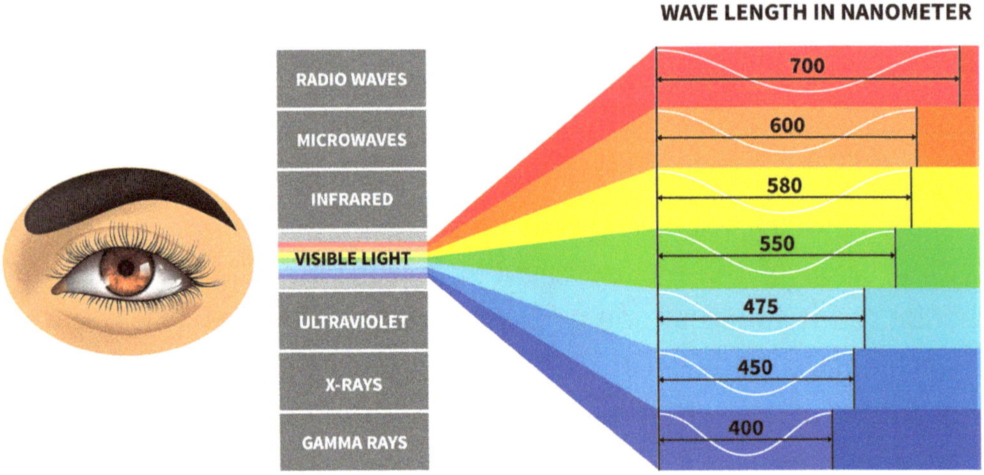

We see the sun as orange-yellow because of how hot it burns and the radiation it gives off. Hotter objects, such as the star Rigel, emit shorter-wavelength radiation, making that star appear blue.

Ultraviolet wavelengths are between 10 nm and approximately 400 nm in length. Because they are shorter, they have a higher frequency (between approximately 7.5×10^{14} and 3×10^{16} hertz).[5] The name for this region of the electromagnetic spectrum, "ultraviolet," comes from the Latin word *ultra*, meaning "beyond." Therefore, the term "ultraviolet" means beyond the violet portion of the visible light spectrum.

More Than a Wave

For hundreds of years, after numerous experiments proving light behaves as a wave, scientists were generally in agreement that

THE BLUE OF BLU-RAY

Music, movies, and games haven't always been streamed over the internet. Before most media was streamed, it was stored on optical discs, called compact discs (CDs), digital versatile discs (DVDs), and Blu-ray discs. One of the major differences among these three disc types is the type of laser (and therefore, wavelength) used to read and write data onto the disc. CDs are written and read using 780 nm red light, DVDs use 650 nm red light, and Blu-ray discs use 405 nm blue-violet light. Because Blu-ray discs use a shorter wavelength, Blu-ray players are able to read the stored data much more precisely, making each disc capable of storing five times more data than DVDs.[6]

light is indeed a wave. This was despite mathematician Sir Isaac Newton's experiments that showed light traveling as particles (which he called corpuscles). Other scientists kept disagreeing with him and trying to prove him wrong.[7] However, in 1905, physicist Albert Einstein proposed that light isn't made of waves or particles—it's made of both. He proposed this theory after multiple experiments with the photoelectric effect in an effort to explain what others had witnessed and couldn't explain.

Scientists knew that shining a strong-enough light on a metal object would give off electrons—this is the photoelectric effect. They hypothesized that if the frequency and the amplitude of the light increased, more electrons would be given off. However, the experiment did not prove this hypothesis. Increasing the frequency didn't change the number of electrons; instead, it changed the

In 1905, Albert Einstein proved light could behave as a particle and a wave. This knowledge eventually led to the creation of solar panels, which use light energy to generate electricity.

amount of energy contained in the electrons given off. This was the opposite of what they expected would happen![8] Their discovery caused a dramatic shift in thinking because it meant that, depending on the situation, light behaves as a wave or as a particle.

This tiny "packet" of energy had many names over the years, from corpuscle to quantum, but by 1926, scientists settled on the term "photon." Think of photons as the cars of a railroad train: Although each car is separate, they all come together to make up a larger train. In the case of light, this train of photons is a stream of energy. Einstein was able to prove that only certain strong-enough frequencies of light (such as ultraviolet light) will cause the photoelectric effect. Because of his work on this theory and the photoelectric effect, Einstein was awarded the 1921 Nobel Prize in Physics.

HISTORIC DISCOVERIES OF INFRARED AND ULTRAVIOLET RADIATION

CHAPTER TWO

Scientists began making the connections between visible white light and all the colors of the rainbow as early as the thirteenth century. It wasn't until Sir Isaac Newton's careful experimentation with prisms and sunlight in the 1660s, though, that scientists finally had evidence that sunlight could be split, or refracted, into a rainbow.[1]

Before Newton's experiment, people thought colors were made of a mix of light and darkness, which were at the ends of this theory's color spectrum. Red, for example, had just a touch of darkness added to the light to make the color. Blue required much more darkness, almost no light, and was the last color on this spectrum before black.[2] Newton challenged this theory and conducted what he called his "crucial experiment."[3] What proved the previous theory wrong was that Newton refracted light through not one prism but two, showing

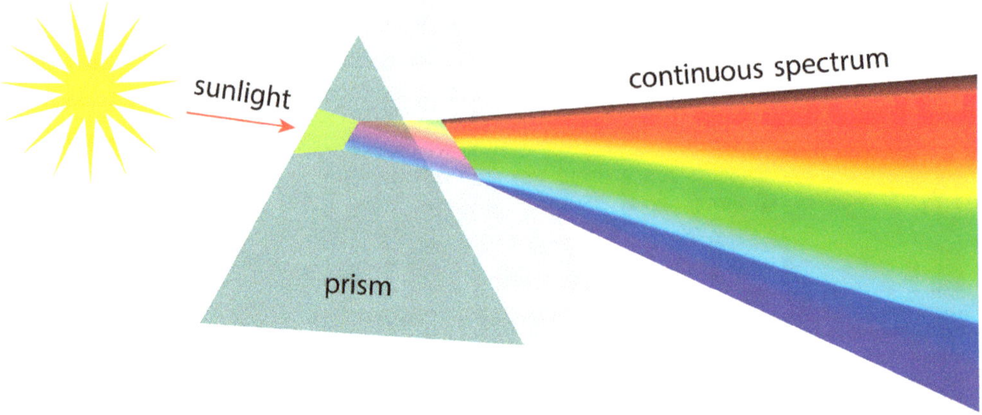

A raindrop acts very similarly to a glass prism. Raindrops higher in the sky bounce red light, while lower ones bounce violet.

that when the rainbow passed through a second prism, it came back together as one beam of white light. After this discovery, it took over a hundred years for scientists to start uncovering other sections of the electromagnetic spectrum.

Discovering the Infrared Spectrum

Although he trained as a professional musician and composer, Sir William Herschel had other ambitions beyond a career in music. He was keenly interested in studying the night sky and building his own telescopes to do just that. Aided by his brother Alexander and his sister Caroline, he made a startling discovery one night in 1781 while looking at the night sky: Herschel discovered the planet Uranus. This skyrocketed him from an

SUNLIGHT AND PRISMS

A prism is a solid but transparent object, usually made of glass, whose sides are cut at specific angles to best refract light. Sunlight contains visible light, ultraviolet, and infrared. When it shines onto a prism, the wavelengths of light bend at different angles. A rainbow appears on the other side of the prism. (Even with a prism, the human eye sees only visible light.) The rainbow appears because the longer wavelengths, like the color red, bend less than the shorter wavelengths, like the color violet. Imagine you have a hand of playing cards, and then you fan them out to see each card in your hand. A prism fans out the colors of visible light. Prisms have also allowed scientists such as Sir William Herschel and Johann Wilhelm Ritter to measure parts of the electromagnetic spectrum separately.[4]

amateur who enjoyed astronomy to one of Britain's leading professional astronomers.

Herschel used different telescopes to view the night sky and the sun. He experimented with ways to filter the sometimes bright and intense light. While experimenting, he noted that different-colored glass lenses did a better or worse job filtering out either light or heat from the objects he was studying. Herschel wrote in his paper from 1800:

> In a variety of experiments I have occasionally made, relating to the method of viewing the sun, with large telescopes, to the best advantage, I used various combinations of differently-coloured darkening glasses. What appeared remarkable was, that when I used some of them, I felt a

> sensation of heat, though I had but little light; while others gave me much light, with scarce any sensation of heat.[5]

This observation led him to a question: Could he detect the difference in temperature of light after it was scattered through a prism?[6] In 1800, Herschel set up this experiment. He filtered sunlight through a prism and onto a number of thermometers lined up to measure the temperatures of the colors on the visible light spectrum. To his delight, the experiment worked as he thought it would! Not only was he able to detect a difference in temperature depending on the color, but he also saw a pattern of temperature increases from violet to red. Even more exciting, he correctly formed the opinion that infrared radiation may exist. He wrote:

> I likewise conclude, that the full red falls still short of the maximum of heat; which perhaps lies even a little beyond visible refraction ... In this case, radiant heat will at least partly, if not chiefly, consist, if I may be permitted the expression, of invisible light.[7]

William Herschel (1738–1822) demonstrated for the first time that there were types of light the human eye could not see.

Herschel went on to conduct follow-up experiments. He measured the temperature beyond the red color bands. Sure enough, he detected a temperature increase caused by infrared light. He did not use the term "infrared;" he opted to call them "calorific rays." His experiment was the first to demonstrate radiant energy outside the visible-light spectrum. He even went so far as to suggest that human eyes were not adapted to see this "invisible" light, which we now know is a limitation of the wavelength-specific cone-cell receptors.

Discovering Ultraviolet Radiation

News of Herschel's infrared discovery quickly spread around the scientific community. German physicist Johann Wilhelm Ritter soon formed the hypothesis that if there is something beyond the red end of the visible-light spectrum, then there surely must be something past the violet end as well. Ritter's background in electrochemistry helped him devise an experiment to test his theory. He knew a certain material, called silver chloride, reacted to sunlight by becoming dark when exposed to it. In an 1801 experiment, Ritter exposed some silver chloride to sunlight refracted by a prism. He saw that where the violet light hit, the silver chloride darkened the quickest, and where the red hit, it darkened more slowly.

Ritter took his experiment further and placed some silver chloride in the darkness next to the violet light. The material turned dark rapidly, despite the absence of visible light. This demonstrated for the first time the existence of light beyond the violet end of the visible-light spectrum. Ritter called this light "chemical rays," though it was later changed to "ultraviolet." He would go on to work with electricity, making batteries, specifically, before he died at the young age of just thirty-three years old.

Unlocking the Full Electromagnetic Spectrum

Herschel and Ritter led the way in proving the existence of light beyond what human eyes can see. More discoveries soon followed. By 1867, James Clerk Maxwell predicted that wavelengths beyond infrared could exist. In 1887, Heinrich Hertz produced radio waves in his laboratory, proving Maxwell correct. Wavelengths shorter than ultraviolet were a bit trickier for scientists to discover and prove. But sure enough, in 1895 Wilhelm Conrad Röntgen accidentally took the first X-ray of his wife's hand.

CHAPTER THREE

INFRARED AND ULTRAVIOLET RADIATION CHARACTERISTICS

Once scientists had discovered these hidden bands of light, they tried to understand the various properties of infrared and ultraviolet wavelengths. They soon realized all forms of light in the electromagnetic spectrum share a few properties. The first—and most important—is that they all travel at the same speed. This speed, which is most appropriately called the speed of light, is considered a constant value in the universe. The speed of light is 186,282 miles (299,792 kilometers) per second. This is incredibly fast; in fact, it's theoretically the fastest speed in the entire universe. If you were in space moving as fast as light travels, you could circle Earth 7.5 times in a single second.[1]

The second important characteristic of all forms of light is that it doesn't require a medium through which to travel. Unlike sound waves, which require a substance like air or water to travel through, light can travel through empty space, called a vacuum. This is why the sun is visible during the day and the stars are visible at night—light travels in space from the sun and other stars to our planet.[2] Although all light wavelengths share these two properties in common, each region of the electromagnetic spectrum has its own unique characteristics.

Different regions of the electromagnetic spectrum also share similar characteristics. These shared traits can be split into two main categories: ionizing and non-ionizing radiation. Ionizing radiation carries a lot more energy and covers the very short wavelengths of gamma rays and X-rays. Large amounts of these types of radiation are very dangerous because wavelengths like these have enough energy to break chemical bonds and affect the electrons of atoms, which are what make up everything, including humans. This is why a dentist puts a heavy lead vest over a patient's torso or a shield over the throat area when taking

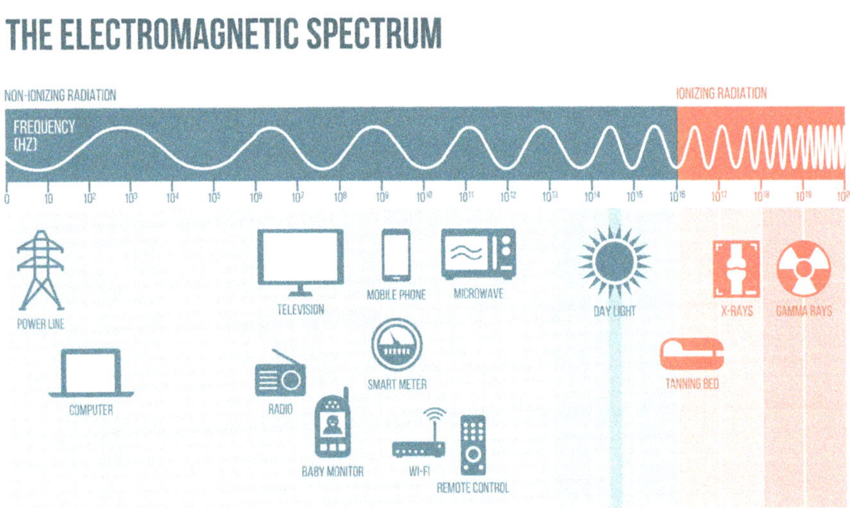

The split between ionizing and non-ionizing radiation occurs in the middle of the ultraviolet section. Longer ultraviolet wavelengths are non-ionizing, but shorter wavelengths are ionizing.

X-rays of the teeth: there's no need to expose the entire body to X-rays when the dentist needs only an X-ray of the mouth area. The vest and shield, rather than the patient's body, absorb the radiation. (It should be noted, though, that X-rays from the dentist alone are likely not enough to cause any real radiation damage.)[3] Fortunately, visible light, infrared, and all wavelengths longer than infrared are non-ionizing, meaning they do not break chemical bonds or pose much of a threat to human health.[4]

Properties of Infrared Radiation

If you've ever sat near a campfire or next to a radiator in your house, you've felt the warmth of infrared radiation. In fact, every object in the universe (except those at the extremely cold temperature of absolute zero, which is –459.67 degrees Fahrenheit, or –273.15 degrees Celsius) emits infrared radiation, including you![5] Infrared radiation is also called thermal radiation. But, somewhat confusingly, not all infrared light carries much heat. For that reason, infrared radiation is usually split into three separate categories: near-infrared (infrared that is closest to visible light), mid-infrared, and far-infrared.

Near-Infrared

Still in the infrared radiation range but closest to the visible-light spectrum is near-infrared. The near-infrared wavelengths are between 0.7 and 1.3 microns long, and they do not provide much heat at all. (A micron is a millionth of a meter.) Most of the infrared light that does make it through Earth's atmosphere is near-infrared.[6]

Mid-Infrared

Farther away from the visible-light spectrum, with wavelengths longer than near-infrared, sit mid-infrared waves. These wavelengths are about 1.3 to 3.0 microns in length. These contain a bit more heat than near-infrared.[7] Mid-infrared light is largely undetectable on Earth's surface because it is filtered out by the atmosphere.

Far-Infrared

The infrared radiation that contains the most heat is far-infrared, which is sometimes called thermal-infrared. This is also the

Thermographic cameras create images like this one, where the hottest objects, such as human bodies, are displayed in white and light colors, and cooler objects, such as clothing, bags, and street surfaces, are darker.

largest range of wavelength sizes, from 3.0 microns all the way up to 1,000 microns. Like mid-infrared, far-infrared light is filtered out by Earth's atmosphere. But the human body, while at a normal temperature, emits radiation around 10 microns. That means humans give off far-infrared light.[8]

Properties of Ultraviolet Radiation

If you've ever seen a black light illuminate hidden ink or make your clothes glow bright in the dark, you've witnessed ultraviolet radiation. Like infrared, ultraviolet can be broken into three main categories, called UVA, UVB, and UVC. UVA is considered non-ionizing, meaning it's not very harmful to humans. UVB and UVC, however, are ionizing. They can damage DNA, skin, and bacteria on the human body. Earth's atmosphere filters out most of the UV light emitted by the sun, which gives people some protection.

Earth's atmosphere absorbs ultraviolet radiation due to UV's interaction with oxygen, particularly in the atmosphere's ozone layer. Ozone is made up of three molecules of oxygen bonded together. But oxygen prefers to be in pairs (just two molecules) and is more stable this way than in groups of three. So when energy from ultraviolet radiation collides with the ozone layer, energy is taken from UV, or absorbed, to form more stable oxygen. However, the short-wavelength UV can also break the oxygen pairs, sending single oxygen molecules to join with pairs, in a constant cycle of destroying and creating ozone. Lucky for us, this cycle ensures people have constant protection from a lot of harmful radiation.[9]

HOW SUNSCREEN WORKS

Although Earth's atmosphere shields humans from most of the damaging effects of ultraviolet radiation, people still need to protect themselves from the UV rays that make it through. Sunscreen protects the skin in two ways. First, it reflects UV off the skin before it can penetrate. Materials used to reflect the rays are usually made of inorganic compounds, such as zinc oxide or titanium dioxide, and have a white appearance. Second, organic compounds absorb UV radiation before the skin can absorb it. When picking a sunscreen, choose one that protects against both UVA and UVB, with a sun-protection factor (SPF) of at least 15.[10]

UVA

UVA radiation begins closest to the violet wavelength of the visible-light spectrum. It is the longest of ultraviolet radiation, with wavelengths between 400 nm and 310 nm. This type of ultraviolet is non-ionizing, but it can still affect humans. UVA rays penetrate deep into the skin, deeper than UVB rays, which affect the top layers of skin. Exposure to UVA rays is known to eventually cause skin aging and wrinkles, and over time can lead to certain skin cancers.

UVB

UVB radiation has wavelengths of between 310 nm and 280 nm. This form of radiation is ionizing, which means it definitely damages DNA. UVB is the form of ultraviolet that causes the skin to sunburn, and prolonged exposure can eventually lead to certain skin cancers. The atmosphere filters out most of

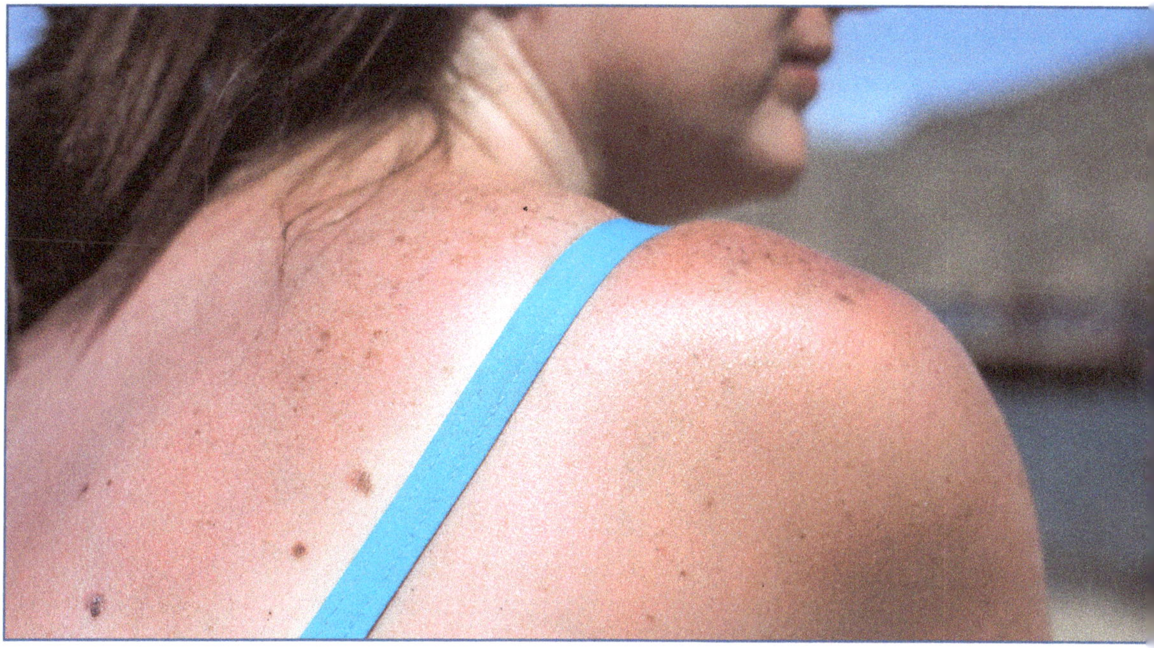

It's important to protect your skin from radiation all year long. Your skin is exposed to direct rays and reflected rays, whether they are bouncing off water in a pool or snow on the ground.

the UVB rays, but even the small percentage of UVB radiation that reaches Earth's surface is dangerous to the skin. Taking precautions against UV radiation, such as wearing sunglasses and protective clothing, avoiding the strongest midday sun, and regularly applying sunscreen, is of paramount importance because it protects the skin from UVA and UVB radiation.

UVC

UVC is the highest-energy form of ultraviolet radiation, and it can be very harmful to biological organisms. The atmosphere filters out all UVC light; it does not reach us. UVC has wavelengths between 280 nm and 100 nm. The shorter the UVC wavelength, the more dangerous it is.

CHAPTER FOUR

STUDYING THE UNIVERSE

Every object in the universe emits light on the electromagnetic spectrum, from massively huge stars to tiny pieces of cosmic dust. Humans have built special telescopes to survey objects like this, whether they be near or far from Earth. These telescopes are specially designed to work with particular wavelengths of light. Some telescopes, such as NASA's Hubble Space Telescope, can see wavelengths from ultraviolet (115 nm) all the way to infrared (2,500 nm). Capturing different wavelengths of light allows scientists to study the surface of suns, the atmosphere of other planets, and the dust and debris that make up planetary rings, like those surrounding Saturn.[1] Others, such as the Galaxy Evolution Explorer (GALEX) telescope, focus strictly on ultraviolet wavelengths, with a mission of finding distant galaxies emitting wavelengths between 135 nm and 280 nm.[2] When designing highly specialized telescopes such as these, engineers must plan very carefully before building anything. The telescopes are extremely expensive, so engineers and scientists can never be

This is an image of the Cygnus Loop Nebula (near the Cygnus constellation). The image was captured by the GALEX telescope, which specifically observes ultraviolet radiation.

too careful. The Hubble Space Telescope, for example, cost NASA $1.5 billion in 1990 when it was built and launched into space.[3]

Before building a telescope, scientists must start with a question: What are they looking for? Are they interested in space as it relates to Earth, or are they researching far-off galaxies? Once they have a firm idea of what they are searching for, they can then answer the next question: What wavelengths of electromagnetic radiation would be best for their research? This answer helps inform the type of telescope they need to build, what data they will need to collect, and where they should put the telescope to collect this data.

Telescope Placement

Because Earth's atmosphere protects people from much of the radiation emitted from objects in space, an interesting problem emerges for scientists hoping to study these waves of light. For example, it's impossible to study a distant galaxy's ultraviolet radiation from here on Earth because of the absorption of most ultraviolet rays by the atmosphere. This problem is addressed by placing telescopes and equipment in one of three main types of observatories, which are buildings specifically dedicated to the study of stars, planets, and even the weather. The first type of observatory is the easiest to build—it's on Earth's surface. The second is a telescope mounted on an airplane that can fly above the ozone layer. The third is a telescope mounted to a satellite that is sent up into space.[4]

Surface Observatories

Humans have been building surface observatories for thousands of years. Structures such as Stonehenge in the United Kingdom, Machu Picchu in Peru, and Angkor Wat in Cambodia are all believed to be ancient observatories used by cultures long gone.

A COMET HITS JUPITER!

Astronomers were treated to an awesome explosion when the Shoemaker-Levy 9 comet broke apart and hit Jupiter in 1994. The event was captured by the Hubble Space Telescope. For the first time, scientists were able to witness two solar system objects colliding. Hubble's strong ultraviolet telescopes detected debris from the collision high in Jupiter's atmosphere, which provided data about the compounds found on the planet and its wind patterns. After this collision, NASA set out to survey all objects that might be a danger to Earth using the Wide-Field Infrared Survey Explorer (WISE) satellite, which uses infrared wavelengths to detect objects.[5]

Sir William Herschel and his sister Caroline, the astronomers who discovered Uranus, built and maintained a facility named Observatory House, which was completed in 1789. Modern-day facilities are state of the art, with millions of dollars' worth of high-tech equipment.

To lessen the interference of Earth's atmosphere, modern surface observatories are often built high up in the mountains. The

La Silla Observatory in Chile is located in the Atacama Desert, which is a popular place for space observation because the skies are incredibly clear almost all year long.

highest of these observatories is the University of Tokyo Atacama Observatory in Chile, which sits atop a mountain 18,504 feet (5,640 meters) above sea level. At this high altitude, a powerful infrared telescope is able to capture wavelengths between 1,000 nm and 2,500 nm.[6]

Aerial Observatories

Scientists realized that in order to study other light wavelengths, including ultraviolet, they needed a way to get around the interference caused by Earth's atmosphere. So they looked toward the sky—and to balloons. By attaching telescopes to durable balloons, scientists were able to not only lift the instruments above the atmospheric interference but also save a lot of money on equipment. The first balloon carrying a telescope and other equipment launched in 1957 as part of the Stratoscope Program. The balloon rose over 80,000 feet (24,384 m). During a few years of missions, scientists gathered thousands of images of the sun's outer layer, called the photosphere. These images, captured in both visible and infrared wavelengths, helped scientists understand the sun's granules. These hot zones of gas make their way from the inside of the sun to the surface.[7]

Another type of aerial observatory uses a modified airplane (a Boeing 747-SP). It is called the Stratospheric Observatory for Infrared Astronomy (SOFIA). In operation since 2010, SOFIA is the world's largest flying observatory ever built. It flies as high as 45,000 feet (13,716 m). This elevation puts it atop 99.9 percent of the water vapor in the atmosphere, which absorbs a lot of infrared radiation. But aircraft present unique challenges, including the motion of flight and the weight restrictions of the

To turn it into SOFIA, this aircraft underwent numerous modifications from 1998 to 2007. Flight testing continued until 2010, when it captured its first images.

instruments. After all, a plane must be able to take off with the telescope on board. Thankfully, the SOFIA aircraft successfully holds a telescope weighing 40,000 pounds (18,144 kilograms) and still has room for scientists, flight crew, and educators to go along for the ride. In 2015, SOFIA was able to fly into the perfect position for the infrared telescopes on board to capture the dwarf planet Pluto as it passed between Earth and a distant star. During this event, called an occultation (or eclipse), infrared radiation emitted from the distant star passed through Pluto's atmosphere. This allowed scientists to collect data about Pluto. Based on the changes in infrared light, scientists learned about the pressure of Pluto's atmosphere, winds on the dwarf planet, and its different atmospheric temperatures.[8]

Satellite Observatories

Although high-mountain surface observatories and aircraft like SOFIA are able to get much-clearer pictures of infrared radiation, the ozone layer still prevents ultraviolet radiation from being studied. To overcome this obstacle, scientists need to get even higher—that is, into space.[9] The first ultraviolet-detecting satellite, the International Ultraviolet Explorer (IUE), was launched in 1978. This satellite provided all sorts of information about the composition of comets, including detecting sulfur on Comet IRAS-Araki-Alcock. Sulfur is believed to be essential for life.

NASA has intentions of increasing its ability to study infrared radiation in space, with the planned launch of the James Webb Space Telescope (JWST). This telescope will be capable of detecting the faintest sources of infrared radiation in the universe, which will detect not only cooler, smaller objects but also extremely distant and older galaxies. The JWST will be so good at detecting infrared light that scientists believe it will be able to find signals from as far back in history as the Big Bang! Because JWST will focus on infrared radiation, scientists have designed a large solar shield to block out the sun's light. This design will keep the telescope cool and avoid interference from the sun. JWST will be the most complex and expensive telescope ever launched.[10]

CHAPTER FIVE

INFRARED AND ULTRAVIOLET RADIATION IN DAILY LIFE

Scientists and astronomers use infrared and ultraviolet radiation to study the universe. But did you know there are a lot of uses for these two forms of light energy much closer to home? In fact, people use them quite frequently every day. A toaster uses infrared to toast bread.[1] And ultraviolet light helps defend against counterfeit money with the addition of an ultraviolet security thread woven into the bills.[2]

Applications Using Infrared

Not all uses of infrared radiation are high-tech. People take advantage of the thermal (or heat) properties of infrared through devices like the simple toaster. Most fast-food establishments use infrared heat lamps to keep meals hot before they're served to customers.[3] Infrared heat lamps are used in the home too. They are commonly used to warm certain pets, such as turtles and other reptiles, as well as baby animals, like chicks and ducklings. (For reptiles though, light bulbs that emit infrared and ultraviolet should be used.) Some homes have infrared lamps in the bathroom, which can keep a person feeling warm even on cold mornings. Saunas also make use of infrared. People (and lots of animals!) love feeling warm, and infrared light bulbs are a safe and effective way to stay cozy.

A very common use of infrared may be sitting in your living room right now, out in the open. Televisions are controlled using infrared (IR) radiation. Built into every IR remote control is a light-emitting diode (LED). This LED pulses infrared signals that are detected by an infrared receiver built into the television. The receiver converts the infrared signal into digital signals that a computer chip can understand. The signals control the televisions power, volume, and channel selection. Some remote controls are stronger than others, which is why you might have to point your remote directly at the television, whereas your friend's remote doesn't have to be aimed quite as carefully.[4] Each manufacturer that produces IR controllers has its own set of commands. Fortunately, universal remotes come preprogrammed with different manufacturer settings, so you can go online or to your neighborhood electronics store and buy one regardless of the brand of television you own.

The military is another industry that uses infrared. Most of the development of infrared technology has occurred during wartime. Militaries around the world try to stay ahead of the enemy, especially with regard to technology that's useful at night. For example, thermal imaging cameras can help soldiers

A remote control uses infrared radiation to signal the receiver in the television.

see using the infrared spectrum when visible light is lacking. The military also uses infrared to guide heat-seeking missiles to their targets. This technology is referred to as heat-seeking because these weapons lock onto the infrared radiation emitted as heat from enemy aircraft. This is extremely helpful because heat-seeking missiles will still hit their targets, even at night. The US Navy was the first to employ infrared heat-seeking missiles when it debuted the AIM-9 Sidewinder missile in the 1950s, during the Vietnam War. The missile gets its name from the sidewinder rattlesnake, which is a pit viper that can see in infrared!

In less life-threatening situations, thermal imaging can improve home-energy efficiency by finding potential heat leaks. It's also used by weather satellites to help detect temperatures of Earth's surface, water, and clouds. IR is a very versatile form of radiation, and humans have developed many interesting ways to take advantage of this wavelength of light.

Ultraviolet Radiation and Fluorescence

Ultraviolet, too, has many common, everyday applications. However, the way humans use ultraviolet radiation isn't quite as straightforward as toasting bread. Because UV can be damaging to the skin, the wavelengths emitted from devices people use need to be carefully controlled. For this

TRY IT OUT!

Modern smartphones are equipped with cameras that are very advanced. In fact, most of them are designed to capture wavelengths between 300 nm and 1,100 nm. This means they pick up not only the full visible-light spectrum but also infrared and ultraviolet. To take better pictures, programmers put in filters to remove any wavelengths outside the visible-light spectrum. But on some phones, you can still see infrared. If you have a smartphone, try holding up the front-facing camera to a remote control. If you press a button on the remote, do you see the remote light on your phone screen?[5]

reason, most of the ways UV light is used are in professional, industrial settings.

One of the properties of ultraviolet light is its ability to cause certain things to fluoresce, or glow, without needing visible light. Fluorescence is an object's ability to give off visible and invisible radiation when in contact with a shorter wavelength, such as ultraviolet. When shorter-wavelength radiation hits an object, lower-energy photons can be given off in the visible-light spectrum, resulting in the ability to see light. The ability of ultraviolet light to create fluorescence is a very powerful property that humans have learned to use in many different ways. Black lights, for example, emit UVA radiation, which is why some things glow, or fluoresce, under black lights.

In the United States and many other countries around the world, the government weaves threads of fluorescent materials into paper money to make it harder to counterfeit. Five-, ten-, twenty-, and fifty-dollar bills each emit a different color, from blue to yellow, when placed under an ultraviolet light. This helps bankers and cashiers quickly check to see whether a bill is real or fake. And fluorescence is not used for security just in money. It is also used in important government-issued documents, such as driver's licenses. Driver's licenses in many states across America have very detailed images hidden from the human eye. Under a UV light, a whole layer of new details emerges.[6]

In 1962, scientists looked to the jellyfish for an exciting breakthrough that would revolutionize biological research. Jellyfish are bioluminescent, meaning that under certain lights, including ultraviolet, they glow. This effect is due to green fluorescent protein (GFP). Once scientists learned how to produce GFP in a lab, they started adding the gene that made GFP into all sorts of organisms, including yeast and mice. To see whether the gene was successfully expressed, scientists placed the test subjects under ultraviolet light in the hopes that they'd glow green—and they did. GFP transformed the biological sciences because researchers now had proof their modifications were working. With the success of GFP, scientists quickly discovered other fluorescent proteins that emit red and orange wavelengths of light (this protein is called red fluorescent protein, or RFP). By using GFP and RFP combined, scientists can mark different genes to see whether they are expressed in the same cells of a single organism. This type of experiment is called co-localization. It is extremely powerful because it helps scientists understand not only the interactions between genes but also where certain genes are located within individual cells.[7]

CHAPTER SIX

USING INFRARED AND ULTRAVIOLET FOR MEDICAL TREATMENTS

Scientists have begun to use ultraviolet and infrared wavelengths of light in the health-care industry. Sometimes IR and UV light is shined directly onto parts of the human body, and other times it is used in a more indirect way to keep people safe from dangerous germs.

Infrared Applications in Health Care

Thermal imaging has been used to help people become more energy efficient in their homes. But it can also create images of the human body! Enter medical infrared thermography: a noninvasive, radiation-free technique that uses, at its most basic, an infrared camera and a computer.[1] The theory behind using infrared imaging is

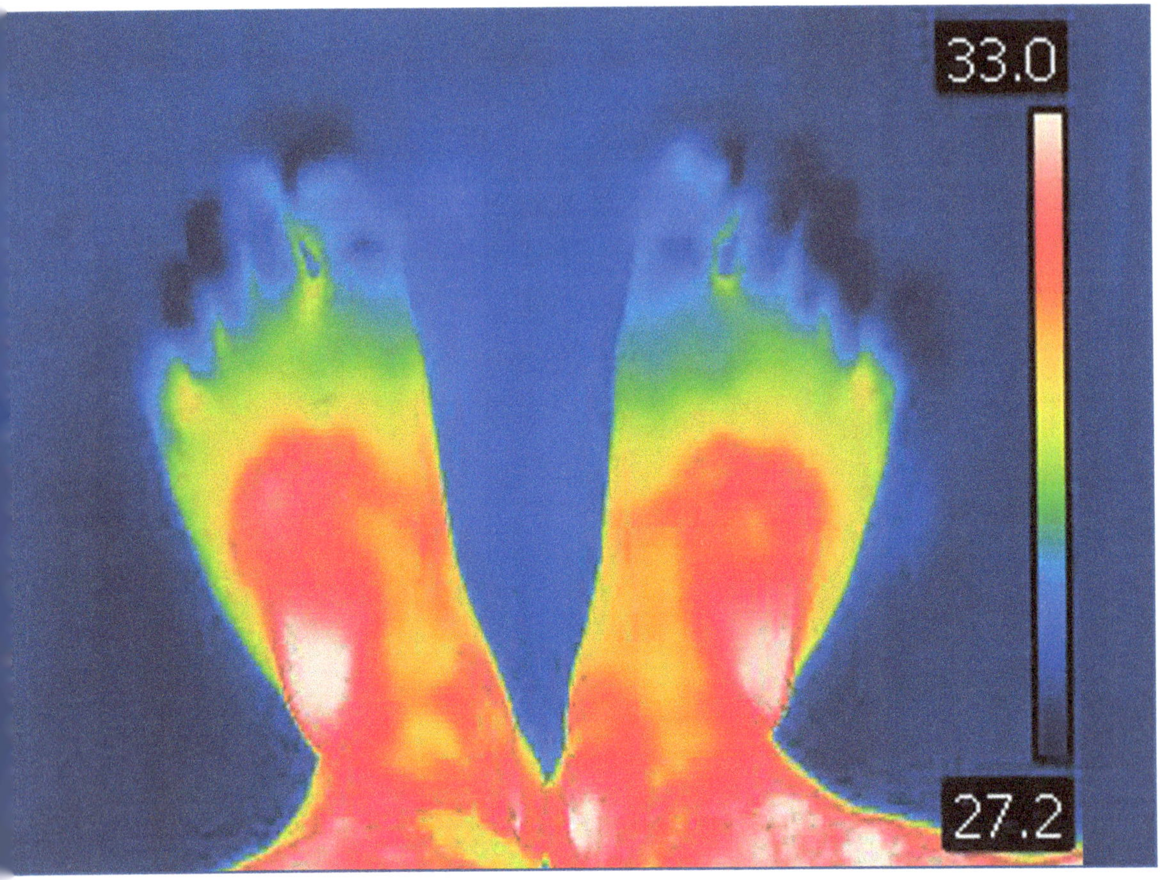

Doctors can use thermal imaging of the human body to detect blood flow issues in patients. People with diabetes often have poor circulation in their feet, which can lead to serious problems.

that subtle differences in body temperature at the surface can indicate fluctuations in blood flow. Changes in blood flow can indicate bigger changes in the body. Researchers have linked injuries such as common knee problems suffered by runners and skiers to differences in blood flow when comparing healthy and injured knees. Infrared imaging is a low-cost, non-ionizing way to examine ailments such as these. But beyond injuries, scientists are also studying whether infrared can accurately diagnosis other

INFRARED FOR FAT LOSS?

People are always looking for quick fixes to expanding waistlines. Infrared light has often been used to promote fat loss, but does it work? The short answer is: we don't know. Available studies on this type of treatment observed only small groups of people, and the studies were not set up scientifically. Most of them did not have a good control group (which is the group that does not receive the experimental treatment—a requirement in any research study), were not long-term (usually spanning only a few months), and did not include long-term follow-ups to see whether the results lasted. For now, at least, it seems the tried and true habits of healthy eating and regular exercise are probably more effective than relaxing in an infrared sauna and expecting to shed excess body fat.[2]

health issues, such as breast cancer, neurological problems, and certain cardiovascular conditions.[3]

Infrared radiation may also directly benefit human health. Because of infrared's ability to transfer heat, it has been used in saunas and localized treatment of various conditions. The thought is that because infrared radiation can penetrate human tissue and muscles without causing damage, it can stimulate blood flow. Improved blood flow carries oxygen to repair damaged tissues; thus, scientists believe infrared treatment could lead to health benefits. Several studies have looked into the use of infrared radiation and have been promising. However, a lack of large-scale studies means the debate over the validity of infrared's purported health benefits continues. Thankfully, because infrared radiation is generally safe to humans, most doctors see no harm in trying.[4]

Ultraviolet Applications for Healthcare

Ultraviolet radiation is more dangerous to humans than infrared because of its higher energy and shorter wavelengths. The safer UVA wavelength is already a staple in everyday lives, but when used appropriately, broad-spectrum ultraviolet light has its benefits, too. In fact, ultraviolet light is frequently used in laboratory and hospital settings as a way to kill germs, keeping patients and experiments from becoming compromised.

It is known that UV light kills bacteria, thanks to the research of Arthur Downes and Thomas P. Blunt, which they published in 1878.[5] By 1895, Doctor Niels Ryberg Finsen was experimenting with using light radiation to treat disease, specifically lupus vulgaris, a form of skin tuberculosis. Although his treatment wasn't a perfect cure, Finsen was awarded the 1903 Nobel Prize in Physiology or Medicine for opening up an entirely new field of medical treatments that uses light radiation.[6]

Tuberculosis (TB) is a very dangerous disease caused by the bacterium *Mycobacterium tuberculosis*, which was first discovered in 1882 by Dr. Robert Koch. It has been infecting humans for thousands of years. At the time, the disease killed one out of every seven people. It was a major epidemic, and there were no antibiotics to treat it. Instead, the most popular treatment doctors prescribed was plenty of fresh air and sunlight (read: exposure to ultraviolet).[7] For decades up until the 1950s, when antibiotics were developed to treat the disease, patients were often moved to sanatoriums. A specific type of hospital, a sanatorium was designed specifically for tuberculosis sufferers: secluded, palatial estates that provided fresh air and sunlight by way of large windows and balconies, both of which were usually facing the south to optimize sun exposure.[8]

The use of UV radiation didn't stop at the treatment of diseases such as TB. After doctors realized the powerful germ-killing properties of ultraviolet radiation, UV lights became standard in hospitals all around the world. UV is used to clean not only surgical equipment and instruments but also entire rooms where sick patients have been staying. The reason UV wavelengths work to sanitize is that exposure to UV damages bacteria's DNA, which makes it impossible for the germs to grow and replicate. The UVC wavelength is particularly effective at killing bacteria because it has the highest energy of the different UV wavelengths. And with the rise of antibiotic-resistant bacteria (meaning the antibiotics that used to fight against bacteria no longer work), the use of ultraviolet radiation for sanitation purposes is becoming more and more important.[9]

Another common place for UV lamps is in science laboratories that work with genetics (i.e., DNA and RNA). In order to prevent cross contamination from sample to sample, technicians work in special stations called laboratory biosafety cabinets, which come with built-in ultraviolet lamps. After each use of the biosafety cabinet, the UV light turns on, decontaminating the workspace for the next project.[10] Meanwhile, passing water under ultraviolet lamps has proven to be another effective method to prevent the

Certain UV lights can be used to kill bacteria, and others can detect it. UV is also used at crime scenes to detect blood and other body fluids.

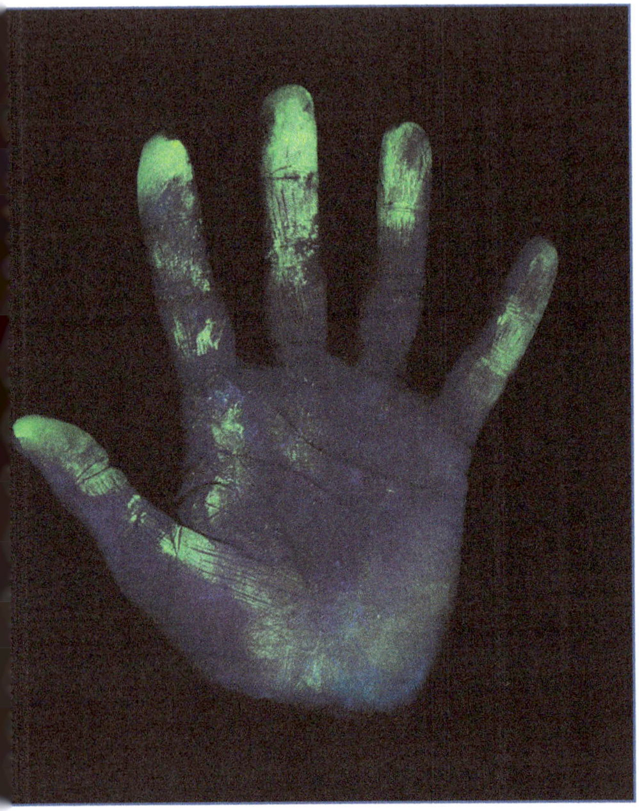

spread of harmful bacteria and viruses.[11]

Though it might seem counterintuitive, doctors in the field of dermatology have also embraced UV light for different medical treatments. Dermatology focuses on the skin, hair, nails, and any conditions affecting those areas of the body. Doctors know that UV radiation is damaging to human skin, but they also know that specific wavelengths of ultraviolet light can treat a variety of skin conditions, including psoriasis, atopic dermatitis, and T-cell lymphoma. One of the most popular "phototherapy" treatments uses a tool that emits a very narrow band of UVB, between the wavelengths of 311 nm and 313 nm. Compare this hyper-targeted UVB radiation to the average tanning booth, which emits a much broader band of UVA and UVB light (320–400 nm and 290–320 nm, respectively).[12] By controlling and focusing the radiation, dermatologists can target the affected skin cells with UVB, which slows their growth.[13]

CHAPTER SEVEN

FUTURE APPLICATIONS

It might feel as though each day is already so full of technologies that use infrared or ultraviolet radiation. From when you make your morning toast to when you use the remote to turn off the television before going to bed, so much of modern-day routines involve the electromagnetic spectrum. And researchers are constantly working to create bigger, better, more innovative uses for technologies that rely on infrared and ultraviolet radiation.

The Future of Infrared

As was discussed, infrared can help people to see in adverse conditions. Engineers are testing whether IR can help self-driving cars see things at any time of day or night. Companies such as Uber and Google have mounted special sensors, called "lidar," onto cars. Similar to radar, the word "lidar" is derived from an acronym that stands for "light detection and ranging," which accurately describes how the technology works. Whereas radar relies on radio waves, lidar devices instead harness the power of infrared light.[1] The lidar

Google's self-driving car uses lidar, which emits a pulse of infrared light and measures how long it takes to come back. The lidar takes these measurements millions of times a second.

system sends out pulses of infrared beams in all directions. When a beam encounters an object, it bounces back to the car. By measuring how long it takes for the beam to return to the car, the lidar system can use the speed of light to calculate how far away objects are and create a detailed map of the car's surroundings. This lets the car determine whether anything is in its way.[2]

Lidar has applications beyond self-driving cars. For example, companies that specialize in creating maps and surveys of land

and water also use lidar systems. One very effective way to map a large area is to attach one of these specialized cameras to an unmanned aerial vehicle (UAV). These vehicles can cover large areas at any time of day without putting a pilot at risk during tricky conditions, such as wildfires. Lidar-equipped UAVs are also excellent for nighttime search-and-rescue missions because the lidar system can identify humans. In less-dangerous situations, lidar-equipped UAVs can survey solar farms and inspect roofs and other construction projects when using a ladder or climbing a structure might not be a safe option.[3]

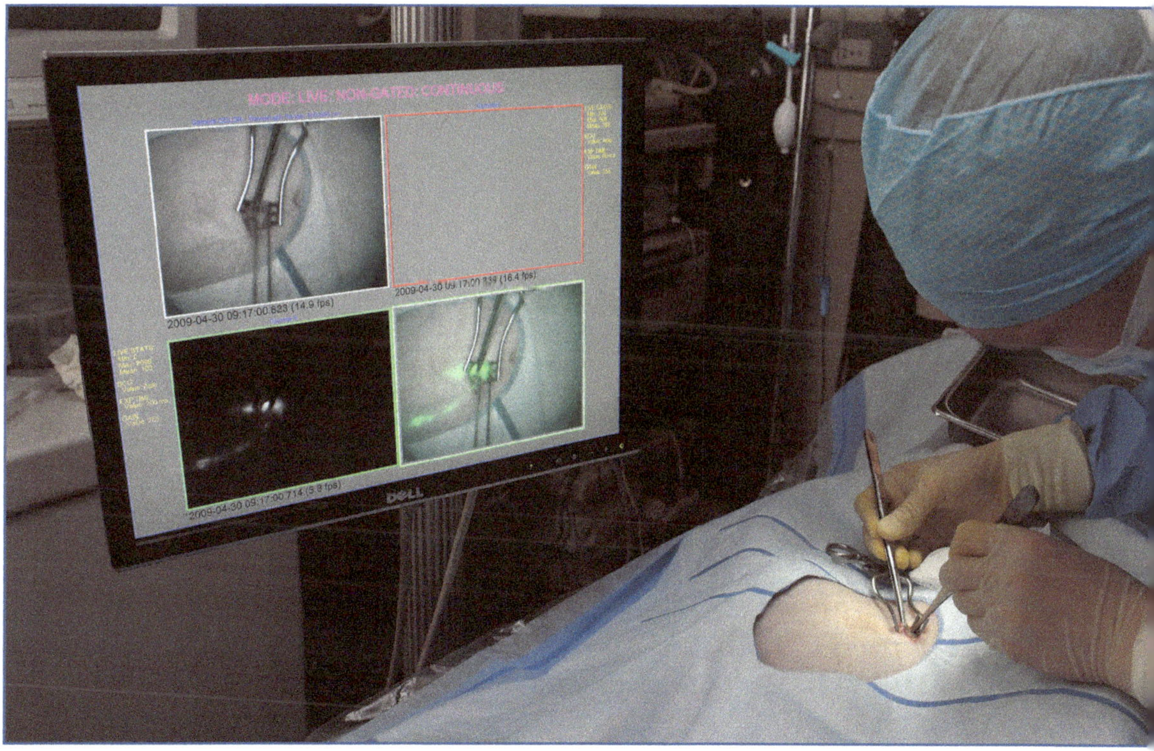

Fluorescence-assisted resection and exploration (FLARE) helps doctors find cancerous cells by causing them to fluoresce (like the green glow on the doctor's screen here).

Use of infrared technology in hospitals has also been improving patient outcomes. In one study of patients after surgery, thermal imaging helped correctly predict whether a bacterial infection was developing over 70 percent of the time.[4] In another example, a team of cancer specialists developed a way to cause tumor cells to give off fluorescence. To do this, they injected volunteer patients with a special dye they had designed to seek out cancerous cells. Doctors then took an infrared image of the patient. The dyes—and therefore, the cancer cells—showed up on the image, allowing doctors to easily locate cancerous cells for removal. This technology is a huge advancement for doctors and has the potential to make it much easier to pinpoint cancer cells. In one patient, the infrared-dye imaging system found two

A LOOK INTO THE BRAIN

Researchers at the University of Alberta in Canada made serious advancements in brain imaging using infrared technology. They figured out a way to detect whether a neuron cell is on or off based on the presence or absence of specific calcium ions. Neurons are specialized cells of the nervous system, and they transmit signals in the brain. In this experiment, when neurons activate, the amount of calcium increases, causing the amount of infrared radiation to decrease. The tool developed by the researchers uses near-infrared light to spot the changes in calcium concentrations. By monitoring the activation and deactivation of neurons, the researchers hope to better understand the brain's ability to learn and create memories, as well as how the brain is affected by stroke.[5]

additional cancerous nodes that had previously been missed by other screening methods.[6]

The Future of Ultraviolet

Advances in technology are not exclusive to the infrared wavelengths of the electromagnetic spectrum. Progress has also been made using ultraviolet wavelengths, especially with the use of a newer source of ultraviolet: light-emitting diodes (LED). LED light bulbs are replacing traditional UV lamps as a more energy-efficient option, and this new technology is capable of emitting specific wavelengths of UV radiation.

Because ultraviolet light damages human skin, being able to target specific wavelengths as safely as possible with LEDs is vital for medical applications. For example, when skin absorbs ultraviolet light, the body produces vitamin D3. But some diseases, such as multiple sclerosis, make it difficult for the body to absorb vitamin D from the foods people eat, and that means exposure to ultraviolet is likely the only way some people can get this vital vitamin. In 2017, a team from Boston University tested to see whether using UVB LEDs would cause the body to produce vitamin D3—and it worked! Patients treated with UVB LED light were on average 2.4 times more efficient at producing vitamin D3 than those who were treated with sunshine alone. And because UVB is potentially harmful, the researchers knew they needed to figure out the optimal wavelength that balanced radiation exposure with health benefits. By targeting such a specific wavelength range, the UVB LED was a more-effective and safer option than prolonged sun exposure.[7]

In the medical world, ultraviolet radiation is perhaps best known for its ability to kill harmful bacteria and viruses. Scientists are working to optimize LED systems to do just that. From sterilizing water to treating food at the grocery store (and even in refrigerators at home), researchers are continually designing new systems to stop harmful bacteria from making people sick—and these new inventions are all made possible by low-power LED.[8] Likewise, advancements in infrared technologies have provided life-saving benefits, from thermal imaging to lidar safety systems in cars. Clearly, infrared applications have come a long way from simply changing the channel on a television. Although people cannot see infrared or ultraviolet radiation with the naked eye, the benefits these "hidden" lights have to offer are all around us, every day.

CHAPTER NOTES

INTRODUCTION

1. Rebecca J. Rosen, "6 Animals That Can See or Glow in Ultraviolet Light," *Atlantic*, August 15, 2011, https://www.theatlantic.com/technology/archive/2011/08/6-animals-that-can-see-or-glow-in-ultraviolet-light/243634.

CHAPTER 1
THE BASICS OF LIGHT ENERGY

1. "What Is the Electromagnetic Spectrum?" HubbleSite, http://hubblesite.org/reference_desk/faq/answer.php.id=70&cat=light.
2. Jim Clark, "Electromagnetic Radiation," Chemguide, 2006, https://www.chemguide.co.uk/analysis/uvvisible/radiation.html.
3. "Facts About Color Blindness," National Eye Institute, February 1, 2015, https://nei.nih.gov/health/color_blindness/facts_about.
4. "Electromagnetic Spectrum," Siyavula, https://www.siyavula.com/read/science/ electromagnetic-grade-10/-radiation/11-electromagnetic-radiation-03.
5. "Electromagnetic Spectrum," Siyavula.
6. Stephanie Watson, "How Blu-Ray Discs Work," HowStuffWorks, October 16, 2004, https://electronics.howstuffworks.com/blu-ray1.htm.
7. Nathaniel Page Stites, "Light I: Particle or Wave?" Visionlearning, 2005, https://www.visionlearning.com/en/library/Physics/24/Light-I/132.

8. "Photoelectric Effect," Khan Academy, https://www.khanacademy.org/science/physics/quantum-physics/photons/a/photoelectric-effect.

CHAPTER 2
HISTORIC DISCOVERIES OF INFRARED AND ULTRAVIOLET RADIATION

1. "The Science of Color," Smithsonian Libraries, https://library.si.edu/exhibition/color-in-a-new-light/science.
2. "Newton and the Color Spectrum," WebExhibits.org, http://www.webexhibits.org/colorart/bh.html.
3. "100 Years Carnegie: Newton—The Crucial Experiment," 100 Years Carnegie Project, Bucknell University, https://www.departments.bucknell.edu/history/carnegie/newton/refraction.html.
4. John Papiewski, "How Do Prisms Work," Sciencing, last updated April 24, 2017, https://sciencing.com/prisms-work-4965588.html.
5. William Herschel, "XIII. Investigation of the Powers of the Prismatic Colours to Heat and Illuminate Objects; with Remarks, That Prove the Different Refrangibility of Radiant Heat," Royal Society Publishing, January 1, 1800, https://royalsocietypublishing.org/doi/10.1098/rstl.1800.0014.
6. "Herschel Discovers Infrared Light," Cool Cosmos, http://coolcosmos.ipac.caltech.edu/cosmic_classroom/classroom_activities/herschel_bio.html.
7. Herschel, "Investigation of the Powers of the Prismatic Colours."

CHAPTER 3
INFRARED AND ULTRAVIOLET RADIATION CHARACTERISTICS

1. Nola Taylor Redd, "How Fast Does Light Travel?" Space, March 7, 2018, https://www.space.com/15830-light-speed.html.
2. "Does Light Travel Forever?" Ask the Van, University of Illinois at Urbana-Champaign, https://van.physics.illinois.edu/qa/listing.php?id=21368&t=does-light-travel-forever.
3. "X-Ray Facility Tips," Vermont Department of Health, http://www.healthvermont.gov/sites/default/files/documents/2016/11/Env_RAD_xray_facility_tips_lead_aprons.pdf.
4. Naseem S., "Be Cautious About Radiation!" Scitable, Nature Education, https://www.nature.com/scitable/blog/our-science/be_cautious_about_radiation_not.
5. "What Is Infrared?" Cool Cosmos, Infrared Processing and Analysis Center, http://coolcosmos.ipac.caltech.edu/cosmic_classroom/ir_tutorial/what_is_ir.html.
6. Margaret Rouse and Jessica Scarpati, "Infrared Radiation (IR)," TechTarget, last updated May 2017, https://searchnetworking.techtarget.com/definition/infrared-radiation.
7. Rouse and Scarpati, "Infrared Radiation (IR)."
8. "The Electromagnetic Spectrum: The Infrared," University of Vienna, https://www.univie.ac.at/geographie/fachdidaktik/FD/site/external_htmls/imagers.gsfc.nasa.gov/ems/infrared.html.
9. "Basic Ozone Layer Science," Environmental Protection Agency, last updated September 24, 2018, https://www.epa.gov/ozone-layer-protection/basic-ozone-layer-science.

10. Kristina Grifantini, "How Does Sunscreen Work?" Live Science, June 25, 2010, https://www.livescience.com/32666-how-does-sunscreen-work.html.

CHAPTER 4
STUDYING THE UNIVERSE

1. Brian Garner, "Hubble Space Telescope," NASA, https://www.nasa.gov/mission_pages/hubble/story/index.html.
2. Mark Wade, "Galex," Astronautix, http://www.astronautix.com/g/galex.html.
3. Nola Taylor Redd, "Hubble Space Telescope: Pictures, Facts & History," Space, December 15, 2017, https://www.space.com/15892-hubble-space-telescope.html.
4. "Observatories Across the Electromagnetic Spectrum," Imagine the Universe! NASA, https://imagine.gsfc.nasa.gov/science/toolbox/emspectrum_observatories1.html.
5. Elizabeth Howell, "Shoemaker-Levy 9: Comet's Impact Left Its Mark on Jupiter," Space, January 24, 2018, https://www.space.com/19855-shoemaker-levy-9.html.
6. Yuzuru Yoshii, Kentaro Motohara, Takashi Miyata, and Natsuko Mitani, "The 1m Telescope at the Atacama Observatory Has Started Scientific Operation, Detecting the Hydrogen Emission Line from the Galactic Center in the Infrared Light," University of Tokyo School of Science, press release, August 11, 2009, https://www.s.u-tokyo.ac.jp/en/press/2009/15.html.
7. Stephen Kidd, "Astronomical Ballooning: The Stratoscope Programme," *New Scientist*, September 17, 1964, pp. 702–704, https://books.google.com/books?id=3qpBF3BjFE4C&pg=PA702#v=onepage&q&f=false; "Perkin Stratoscope

II Telescope Model," National Air and Space Museum Collection, Smithsonian Institution, https://airandspace.si.edu/node/34357.

8. James M. De Buizer, "Early Science Results from SOFIA, the World's Largest Airborne Observatory," Universities Space Research Association—Stratospheric Observatory for Infrared Astronomy, Cornell University, February 20, 2013, https://arxiv.org/pdf/1302.5042.pdf.
9. "FAQs," SOFIA Science Center, https://www.sofia.usra.edu/multimedia/about-sofia/faqs.
10. "About: Webb vs Hubble Telescope," James Webb Space Telescope, National Aeronautics and Space Association, https://jwst.nasa.gov/comparison_about.html#.

CHAPTER 5
INFRARED AND ULTRAVIOLET RADIATION IN DAILY LIFE

1. Marshall Brain, "How Toasters Work," HowStuffWorks, May 3, 2000, https://home.howstuffworks.com/toaster1.htm.
2. "How to Detect Counterfeit US Money with a Royal Sovereign Counterfeit Detector," Royal Sovereign, https://www.royalsovereign.com/shoppers-guide/how-to-detect-counterfeit-us-money-with-a-royal-sovereign-counterfeit-detector.html.
3. Sarah Connor, "The Many Uses of Heat Lamps," Lightbulbs.com, February 15, 2016, https://www.lightbulbs.com/blog/uses-for-heat-lamps.
4. Julia Layton, "How Remote Controls Work," HowStuffWorks, November 10, 2005, https://electronics.howstuffworks.com/remote-control1.htm.

5. "How to See the Invisible Infrared World Using Your Mobile Phone Camera," Ken's Tech Tips, June 27, 2018, https://kenstechtips.com/index.php/how-to-see-the-invisible-infrared-world-using-your-mobile-phone-camera.
6. Leah Pandiscia, "Fluorescence Detection in the Detection of Counterfeit US Currency," JASCO, https://jascoinc.com/applications/using-fluorescence-detection-detection-counterfeit-us-currency.
7. Roger Y. Tsien, "The Green Fluorescent Protein," *Annual Review of Biochemistry* 67 (July 1998): 509–544, https://www.annualreviews.org/doi/10.1146/annurev.biochem.67.1.509.

CHAPTER 6
USING INFRARED AND ULTRAVIOLET FOR MEDICAL TREATMENTS

1. Stephanie Watson, "What Is Thermography," Healthline, https://www.healthline.com/health/breast-cancer/thermography.
2. Danielle Pavliv, Sandy Wang, and Varuna Srinivasan, "Does Infrared Light Therapy Work for Weight Reduction?" National Center for Health Research, http://www.center4research.org/infrared-light-therapy-work-weight-reduction.
3. Carolin Hildebrandt, Christian Raschner, and Kurt Ammer, "An Overview of Recent Application of Medical Infrared Thermography in Sports Medicine in Austria," *Sensors (Basel)* 10, no. 5 (2010): 4,700–4,715, https://www.ncbi.nlm.nih.gov/pmc/articles/PMC3292141/ .
4. Yu-Hsin Lin and Tzong-Shiun Li, "The Application of

Far-Infrared in the Treatment of Wound Healing: A Short Evidence-Based Analysis," *Journal of Evidence-Based Complementary and Alternative Medicine* 22, no. 1 (January 2017): 186–88, https://www.ncbi.nlm.nih.gov/pmc/articles/PMC5871200.

5. Surgically Clean Air, "Ultraviolet Light (UV-C): Highly Effective Way to Kill Germs & Viruses," Surgically Clean Air, https://www.surgicallycleanair.com/ultraviolet-light-uv-c-a-highly-effective-way-to-kill-germs-and-viruses/.
6. "Niels Ryberg Finsen: Biographical," from *Nobel Lectures, Physiology or Medicine 1901*–1921 (Amsterdam, Netherlands: Elsevier Publishing Company, 1967), https://www.nobelprize.org/prizes/medicine/1903/finsen/biographical.
7. "World TB Day 2018" Centers for Disease Control and Prevention, December 12, 2016, https://www.cdc.gov/tb/worldtbday/history.htm.
8. Anthony Denzer, "Tuberculosis and Solar Architecture (Part 2)," *Solar House History* (blog), July 19, 2013, http://solarhousehistory.com/blog/2013/7/19/tb2.9.
9. Debbe Geiger, "UV Light Helps Duke Hospitals Fight Transmission of Superbugs," Duke Health, January 17, 2017, https://www.dukehealth.org/blog/uv-light-helps-duke-hospitals-fight-transmission-of-superbugs.
10. Paul J. Meechan and Christina Wilson, "Use of Ultraviolet Lights in Biological Safety Cabinets: A Contrarian View," *Applied Biosafety* 11, no. 4 (December 2006): 222–27, https://journals.sagepub.com/doi/abs/10.1177/153567600601100412.

11. Kai Song, Fariborz Taghipour, and Madjid Mohseni, "Microorganisms Inactivation by Wavelength Combinations of Ultraviolet Light-Emitting Diodes (UV-LEDs)," *Science of the Total Environment* 665 (May 15, 2019): 1,103–1,110, https://www.ncbi.nlm.nih.gov/pubmed/30893742.
12. "Phototherapy," National Psoriasis Foundation, https://www.psoriasis.org/about-psoriasis/treatments/phototherapy.
13. Blake P. Sampson, Aaron M. Secrest, Christopher B. Hansen, and Allen C. Geller, "Examining Dermatologist Use and Opinions of Ultraviolet Radiation for Cosmetic and Medical Purposes," *Journal of Clinical and Aesthetic Dermatology* 11, no. 2 (February 2018): 41–46, https://www.ncbi.nlm.nih.gov/pmc/articles/PMC5843361.

CHAPTER 7
FUTURE APPLICATIONS

1. Devin Coldewey, "Here's How Uber's Self-Driving Cars Are Supposed to Detect Pedestrians" *TechCrunch*, 2018, https://techcrunch.com/2018/03/19/heres-how-ubers-self-driving-cars-are-supposed-to-detect-pedestrians.
2. NOAA, "What Is Lidar?" National Oceanic and Atmospheric Administration, last updated June 25, 2018, https://oceanservice.noaa.gov/facts/lidar.html.
3. Fintan Corrigan, "12 Top Lidar Sensors for UAVs, Lidar Drones and So Many Great Uses," DroneZon, May 26, 2019, https://www.dronezon.com/learn-about-drones-quadcopters/best-lidar-sensors-for-drones-great-uses-for-lidar-sensors.

4. Charmaine Childs et al., "The Surgical Wound in Infrared: Thermographic Profiles and Early Stage Test-Accuracy to Predict Surgical Site Infection in Obese Women During the First 30 Days After Caesarean Section," *Antimicrobial Resistance and Infection Control* 8, no. 7 (January 7, 2019), https://aricjournal.biomedcentral.com/articles/10.1186/s13756-018-0461-7.
5. Katie Willis, "New Tool Enables Imaging of Neural Activity with Near-Infrared Light," Faculty of Science, University of Alberta, January 21, 2019, https://www.ualberta.ca/science/science-news/2019/january/neuron-activity-detector.
6. Missael Garcia et al., "Bio-inspired Imager Improves Sensitivity in Near-Infrared Fluorescence Image-Guided Surgery," *Optica* 5, no. 4 (2018): 413–22, https://www.ncbi.nlm.nih.gov/pmc/articles/PMC6241325.
7. T. A. Kalajian, A. Aldoukhi, A. J. Veronikis, K. Persons, and M. F. Holick, "Ultraviolet B Light Emitting Diodes (LEDs) Are More Efficient and Effective in Producing Vitamin D3 in Human Skin Compared to Natural Sunlight," *Scientific Reports* 7, no. 11489 (September 13, 2017), https://www.nature.com/articles/s41598-017-11362-2.
8. Tatiana Koutchma, Andrew Green, and Vladimir Popovic, "UV Light-Emitting Diodes: The Upcoming Solution for Enhanced Food Safety," *Food Safety*, August/September 2018, https://www.foodsafetymagazine.com/magazine-archive1/augustseptember-2018/uv-light-emitting-diodes-the-upcoming-solution-for-enhanced-food-safety.

GLOSSARY

ABSORPTION A substance or molecule's ability to take in and capture light energy.

BIOLUMINESCENCE The production and emission of light by a living organism.

CO-LOCALIZATION The overlap of two fluorescent signals that suggests an occurrence of two things together in a cell.

ELECTROMAGNETIC SPECTRUM All differing wavelengths of light energy, from gamma rays to radio waves.

EXPRESSED When a specific gene is activated in a living organism.

FILTERS Materials that block or reduce specific wavelengths of light.

FREQUENCY The number of crests of a wave that pass a certain point in a given amount of time. Frequency is measured in hertz, or crests per second.

INFRARED RADIATION Light energy made up of wavelengths between 750 nanometers and 1 millimeter.

INVERSE Something that is the opposite of something else.

IONIZING When radiation has enough energy to detach electrons from an atom or a molecule.

MEDIUM The substance in which something lives or acts.

MICRON A unit of length equal to one millionth (10^{-6}) of a meter.

NANOMETER A unit of length equal to one billionth (10^{-9}) of a meter.

NON-IONIZING When radiation does not have enough energy to detach electrons from an atom or a molecule.

OZONE LAYER A layer of Earth's atmosphere that absorbs most of the sun's ultraviolet radiation.

PHOTOELECTRIC EFFECT The emission of electrons when light energy hits certain surfaces.

PHOTON A single particle of electromagnetic radiation or light.

PHOTORECEPTOR CELLS Specialized cells in an organism that perceive light energy.

RADIATION The process of giving off energy as waves or particles.

REFRACTION The bending of light when it passes at an angle from one medium into another (as when light passes from air into water).

STERILE Free from living organisms, especially pathogens and microbes.

ULTRAVIOLET RADIATION Light energy made up of wavelengths between 10 nanometers and 400 nanometers.

VISIBLE LIGHT Light energy made up of wavelengths between 400 nanometers and 750 nanometers.

WAVELENGTH The distance between one crest of a wave and the next.

FURTHER READING

BOOKS

Haelle, Tara. *Edible Sunlight*. Vero Beach, FL: Rourke Pub Group, 2016.

Kruesi, Liz, and David A. Weintraub. *Astronomy*. Minneapolis, MN: Essential Library, 2016.

Latta, Sara L. *Black Holes: The Weird Science of the Most Mysterious Objects in the Universe.* Minneapolis, MN: Twenty-First Century Books, 2018.

Royston, Angela. *All About Light.* Chicago, IL: Heinemann Raintree, 2016.

Sirota, Lyn. *Information Waves.* Vero Beach, FL: Rourke Pub Group, 2016.

WEBSITES

Hubblesite.org

hubblesite.org/explore_astronomy

This website is completely dedicated to the Hubble Space Telescope, with rich images and additional information about the mission.

PBS Learning Media (Requires FLASH)

ny.pbslearningmedia.org/asset/phy03_int_spectrum/

Take an interactive tour of the full electromagnetic spectrum.

University of Colorado Boulder PhET Interactive Simulator

phet.colorado.edu/en/simulations/category/physics/light-and-radiation

This interactive website allows you to explore the different properties of light radiation.

INDEX

N

R

S

T

U

V

W

X